INTRODUCTION TO SLIME

INTRODUCTION

Slime is a fun and squishy material that has become incredibly popular in recent years. It's not only fun to play with, but it's also a great way to learn about chemistry and physics. Slime can be made in many different ways, each with its own unique properties and characteristics.

Types of Slime

The most common types of slime are clear slime, fluffy slime, and butter slime. Clear slime is a transparent slime that has a smooth and shiny appearance. Fluffy slime is a light and airy slime that is easy to stretch and mold. Butter slime is a soft and moldable slime that is usually made with clay.

In addition to these types of slime, there are many other variations of slime that can be made, including glitter slime, magnetic slime, and glow-in-the-dark slime. Each type of slime has its own unique texture and characteristics.

Slime is Good for Sensory

Slime is a tactile material that can provide a unique sensory experience for individuals of all ages. It can offer a range of textures, colors, and scents to explore, helping to stimulate the senses and promote relaxation. Additionally, playing with slime can encourage creativity, imagination, and problem-solving skills.

Basic Steps for Making Slime

While the specific steps for making slime will vary depending on the type of slime you want to make, there are some basic steps that apply to most types of slime:

Mix the glue and any other desired add-ins in a bowl.

Add the activator (such as borax or saline solution) to the glue mixture and stir until the slime forms.

Knead the slime until it becomes smooth and stretchy.

If desired, add food coloring or glitter to the slime and mix until the desired color is achieved.

Making slime is a fun and creative activity that can be enjoyed by people of all ages. In the next chapter, we'll dive into the science behind slime and how it works.

THE SCIENCE BEHIND SLIME

Slime is not only a fun and squishy material to play with, but it is also a great way to learn about the science behind it. We'll dive into the science behind slime, including its properties and how it is made.

Properties of Slime

Slime has several unique properties that make it an interesting material to study. Two of the most important properties of slime are viscosity and elasticity.

Viscosity refers to a substance's resistance to flow. Slime has a high viscosity, which means it has a thick and gooey consistency. The more viscous a substance is, the more it resists flowing.

Elasticity refers to a substance's ability to stretch and return to its original shape. Slime is highly elastic, which means it can be stretched and pulled without breaking. The more elastic a substance is, the more it can be stretched without deforming or breaking.

Slime is a non-Newtonian fluid, which means it doesn't follow the typical rules of fluid dynamics. Unlike a Newtonian fluid, such as water, the viscosity of slime changes depending on the amount of stress applied to it. When slime is at rest, it behaves like a liquid, but when you pull or stretch it, it becomes more solid.

Other properties of slime include its color, transparency, and scent. These properties can be altered by adding different ingredients such as food coloring or essential oils.

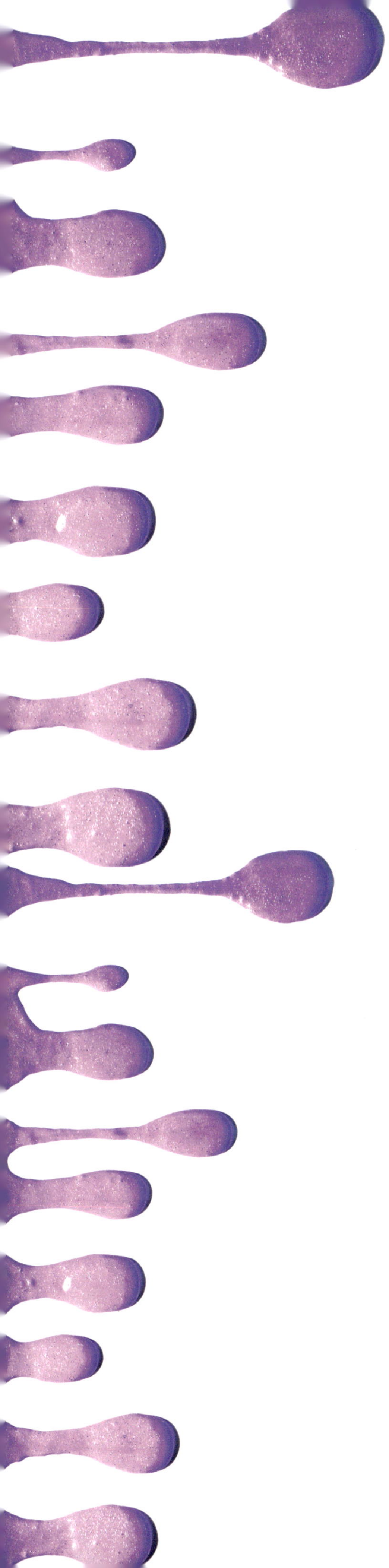

Science Behind Making Slime

Slime is a fun and interesting substance that can teach us about science! Slime is a viscous, stretchy substance that can be made using a variety of ingredients, including glue, borax, laundry detergent, and cornstarch. Slime is made by cross-linking long chains of molecules called polymers. This process is called polymerization, and it happens when certain chemicals react and link together. The more cross-links between the polymers, the more viscous, or thick and gooey, the slime becomes. So, the more polymerization that occurs, the thicker and stretchier the slime gets!

Slime is made through a chemical reaction between two main ingredients: glue and an activator. Glue is an important ingredient in slime because it contains a polymer called polyvinyl acetate (PVA). PVA glue contains long chains of molecules called polymers. When mixed with other ingredients (activators) like borax or contact lens solution, the reaction causes the polymer chains to link together (cross-linking), forming a network of molecules. This network of molecules creates the squishy and stretchy texture of slime.
The most common activators used in slime-making are borax and saline solution. When mixed with the glue, these activators react with the polymer chains to form the slime. Borax works by breaking down the bonds between the polymer chains, allowing them to link together and form slime. The saline solution works in a similar way, but it contains sodium ions that also help to link the polymer chains.

Slime is a fascinating material that has many unique properties and characteristics. By understanding the science behind slime, we can learn about important scientific concepts such as polymers, viscosity, and elasticity. In the next chapter, we'll explore the basic slime recipes and how to make them.

BASIC TYPES OF SLIME

Special Note: Be sure to have large mixing bowls, measuring cups/spoons, something to mix with, and a tablecloth for your table. An apron or play clothes are appropriate. Gloves are optional.

INGREDIENTS

- **White or clear PVA Washable School Glue**
- **Saline Solution**
- **Borax Powder**
- **Liquid Starch**
- **Baking Soda**
- **Water**
- **Food Coloring and Fun Mix-Ins**

INGREDIENTS

- White or clear PVA Washable School Glue
- Saline Solution
- Borax Powder
- Liquid Starch
- Baking Soda
- Water
- Food Coloring and Fun Mix-Ins

SALINE SOLUTION SLIME

1. Mix 1/2 cup of glue and 1/2 cup of water in a bowl.
2. Add 1/2 tsp of baking soda and stir.
3. Add food coloring and mix-ins if desired.
4. Add 1 tbsp of saline solution.
5. Stir vigorously until mixture forms slimy ball.

Coat hands with a few drops of saline and knead slime to desired consistency.

LIQUID STARCH SLIME

1. Mix 1/2 cup of glue and 1/2 cup of water in a bowl.
2. Add food coloring and mix-ins if desired.
3. Add 1/4 cup of liquid starch
4. Stir vigorously until mixture forms slimy ball.
5. Knead slime with hands and let it rest
6. for 3 minutes or knead immediately.

Slime may appear stringy at first but the texture will change as you knead it.

BORAX SLIME

1. Mix 1/2 cup of glue and 1/2 cup of water in a bowl.
2. Add food coloring and mix-ins if desired.
3. In a separate container combine 1/4 tsp of borax powder and 1/2 cup of warm water and stir well.
4. Add borax solution to glue mixture and stir
5. vigorously until mixture forms a slimy ball

Knead slime with hands for desired consistency.

CLEAR SLIME

INGREDIENTS

1/2 cup clear PVA glue

1/2 cup water

1/2 tsp baking soda

1 tbsp contact lens solution

Clear slime is a popular type of slime that is transparent and has a smooth and shiny appearance. Here's how to make clear slime:

DIRECTIONS:

In a bowl, mix together the clear glue and water until well combined.

Add the baking soda and mix well.

Add the contact lens solution and stir until the slime forms and pulls away from the bowl.

Knead the slime until it becomes smooth and stretchy.

The science behind clear slime: Clear slime is made by mixing clear PVA glue with an activator, which is typically a contact lens solution. The activator reacts with the polymer chains in the glue, causing them to link together and form the slime.

FLUFFY SLIME

INGREDIENTS

1/2 cup PVA glue

1/2 cup shaving cream

1/2 tsp baking soda

1 tbsp contact lens solution

BACKGROUND

Fluffy slime is a light and airy type of slime that is easy to stretch and mold. Here's how to make fluffy slime:

DIRECTIONS:

In a bowl, mix together the glue and shaving cream until well combined.

Add the baking soda and mix well.

Add the contact lens solution and stir until the slime forms and pulls away from the bowl.

Knead the slime until it becomes smooth and stretchy.

The science behind fluffy slime: Fluffy slime is made by adding shaving cream to the basic slime recipe. The shaving cream adds air to the slime, which makes it lighter and more fluffy.

BUTTER SLIME

BACKGROUND

Butter slime is a soft and moldable type of slime that is usually made with clay. Here's how to make butter slime:

INGREDIENTS:

1/2 cup PVA glue
1/2 cup water
1/2 tsp baking soda
1 tbsp contact lens solution
1/4 cup model magic clay

DIRECTIONS:

In a bowl, mix together the glue and water until well combined.

Add the baking soda and mix well.

Add the contact lens solution and stir until the slime forms and pulls away from the bowl.

Knead the slime until it becomes smooth.

Add the model magic clay to the slime and knead until it is fully incorporated.

The science behind butter slime: Butter slime is made by adding a soft and moldable ingredient such as clay to the basic slime recipe. The clay adds texture and makes the slime more moldable.

CONCLUSION:

These basic slime recipes are a great starting point for exploring the world of slime. By understanding the science behind each type of slime, we can learn about important scientific concepts such as polymers and chemical reactions. In the next chapter, we'll explore specialty slime recipes that use unique ingredients to create even more exciting types of slime.

SPECIALTY SLIME

Welcome to the specialty slime section! If you're looking to take your slime-making skills to the next level and create something truly unique and exciting, you've come to the right place. In this section, we will explore some of the most fascinating and innovative slime recipes that will capture your imagination and spark your creativity.

Gone are the days of simple, run-of-the-mill slime. In this section, we will introduce you to a whole new world of slime making, where glitter, foam beads, and other fun additions can be used to create a variety of textures and effects. From crunchy and fluffy slime to sparkly and glow-in-the-dark slime, we've got it all covered.

Looking for a slime that will give you a satisfying crunch when you squish it? Try making crunchy slime by adding foam beads or other crunchy elements to your slime mixture. Want to make a fluffy, cloud-like slime that feels like you're playing with a pillow? You can create a super soft and airy texture by using shaving cream or instant snow in your slime recipe.

If you're in the mood for something sparkly and glittery, we have just the thing for you. Glitter slime is a classic favorite and can be made in a variety of colors and styles. But why stop at glitter when you can create a whole galaxy in your slime? With glow-in-the-dark slime, you can make a slime that lights up the night and captures the beauty of the universe.

SPECIALTY SLIME

And that's not all - we'll also introduce you to butter slime, jiggly slime, and even edible slime! There's something for everyone in this specialty slime section, whether you're a beginner or an experienced slime maker. So, get ready to unleash your creativity and have some fun exploring the world of specialty slimes!

GLOW IN THE DARK SLIME

Glow-in-the-dark slime is a fun and unique type of slime that glows in the dark. Here's how to make it:

INGREDIENTS

1/2 cup PVA glue
1/2 cup water
1/2 tsp baking soda
1 tbsp contact lens solution
1 tsp phosphorescent powder

DIRECTIONS:

In a bowl, mix together the glue and water until well combined.
Add the baking soda and mix well.

Add the phosphorescent powder and mix until well incorporated.
Add the contact lens solution and stir until the slime forms and pulls away from the bowl.

Knead the slime until it becomes smooth and stretchy.

Charge the slime under a light for several minutes, then turn off the lights to see it glow.

The science behind glow-in-the-dark slime: The phosphorescent powder in glow-in-the-dark slime absorbs light energy and then releases it as light over time. This is known as phosphorescence. When the slime is exposed to light, the phosphorescent powder absorbs the energy and stores it. When the lights are turned off, the powder releases the stored energy as light, causing the slime to glow.

MAGNETIC SLIME

MAGNETIC SLIME

Magnetic slime is a type of slime that can be manipulated with a magnet. Here's how to make it:

DIRECTIONS:

In a bowl, mix together the glue, iron oxide powder, and water until well combined.
Add the baking soda and mix well.
Add the contact lens solution and stir until the slime forms and pulls away from the bowl.

Knead the slime until it becomes smooth and stretchy.

Use the magnet to manipulate the slime.
The science behind magnetic slime: Iron oxide powder is a magnetic material that can be attracted to a magnet. When it is mixed with the glue, it creates a magnetic slime that can be manipulated with a magnet.

INGREDIENTS

1/2 cup PVA glue

1/2 cup iron oxide powder

1/2 cup water

1/2 tsp baking soda

1 tbsp contact lens solution

Neodymium magnet

OOBLECK SLIME

Oobleck and other substances (such as Silly Putty and quicksand) are not liquids such as water or oil. They are known as non-Newtonian fluids. That is, it acts like a liquid when being poured, but like a solid when a force is acting on it. You can grab it and then it will ooze out of your hands. Make enough Oobleck and you can even walk on it!

Oobleck gets its name from the Dr. Seuss book Bartholomew and the Oobleck where a gooey green substance, Oobleck, fell from the sky and wreaked havoc in the kingdom.

Most of it has to do with pressure. The size, shape, and makeup of the cornstarch grains causes the cornstarch to "lock-up" and hold its shape when pressure is applied to it.

DIRECTIONS

1. Pour the cornstarch into the bowl. Don't rush to add water.
2. Add the food coloring to your water before adding it to the powder.
3. Add water. There are no exact formulas regarding how much water to add, but it will end up being about 1/2 cup (120 ml) of water per cup (235 ml) of cornstarch. The secret is to add the water slowly and mix as you add it.
4. Dig in with your hands and really mix it up

INGREDIENTS

- **1-2 cups of Cornstarch**
- **1/2 - 1 cup of Water**
- **Food coloring (we always say it's optional, but it does make it more fun - don't use too much or you could end up with colored hands, clothes...etc.)**

OOBLECK

BACKGROUND

Is it a solid? Is it a liquid? Can it be both? Play around with it and find out how it acts differently from a normal liquid and a normal solid. Applying pressure to the mixture increases its viscosity (thickness). A quick tap on the surface of Oobleck will make it feel hard, because it forces the cornstarch particles together. But dip your hand slowly into the mix, and see what happens—your fingers slide in as easily as through water. Moving slowly gives the cornstarch particles time to move out of the way.

Glitter Slime

BACKGROUND

Glitter slime is a fun and sparkly type of slime that is perfect for kids. Here's how to make it:

INGREDIENTS:

1/2 cup PVA glue
1/2 cup water
1/2 tsp baking soda
1 tbsp contact lens solution

DIRECTIONS:

In a bowl, mix together the glue and water until well combined.

Add the baking soda and mix well.

Add the glitter and mix until well incorporated.

Add the contact lens solution and stir until the slime forms and pulls away from the bowl.

Knead the slime until it becomes smooth and stretchy.

The science behind glitter slime: Glitter is a fun and sparkly addition to slime that doesn't have much of a scientific explanation. It's just a fun and creative way to customize your slime.

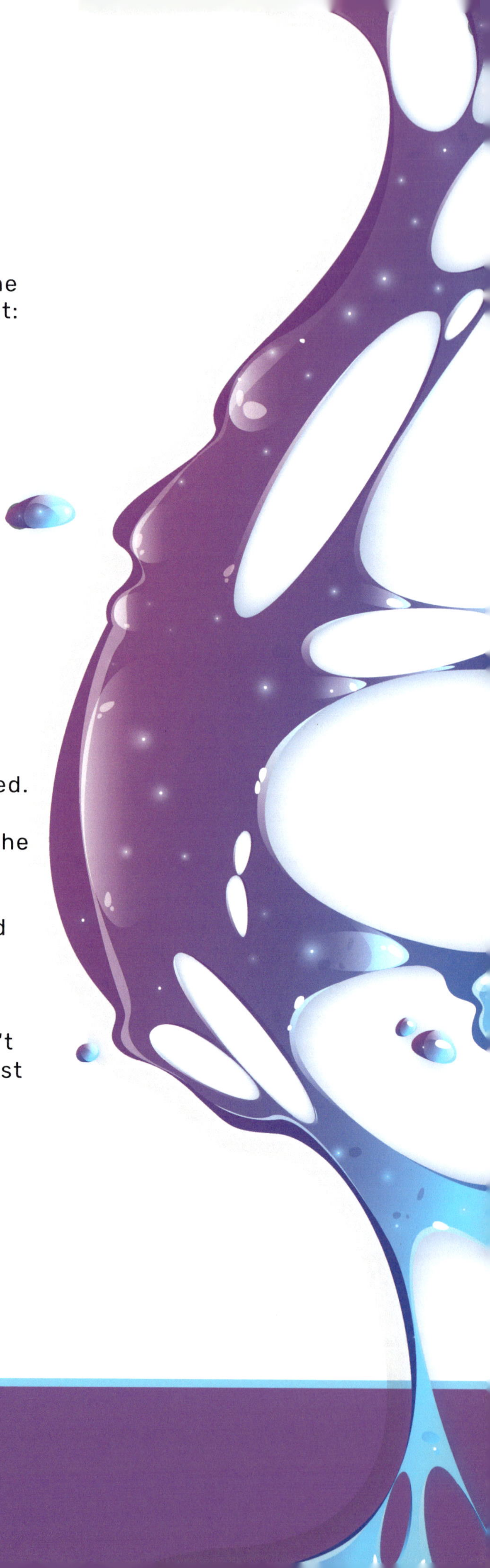

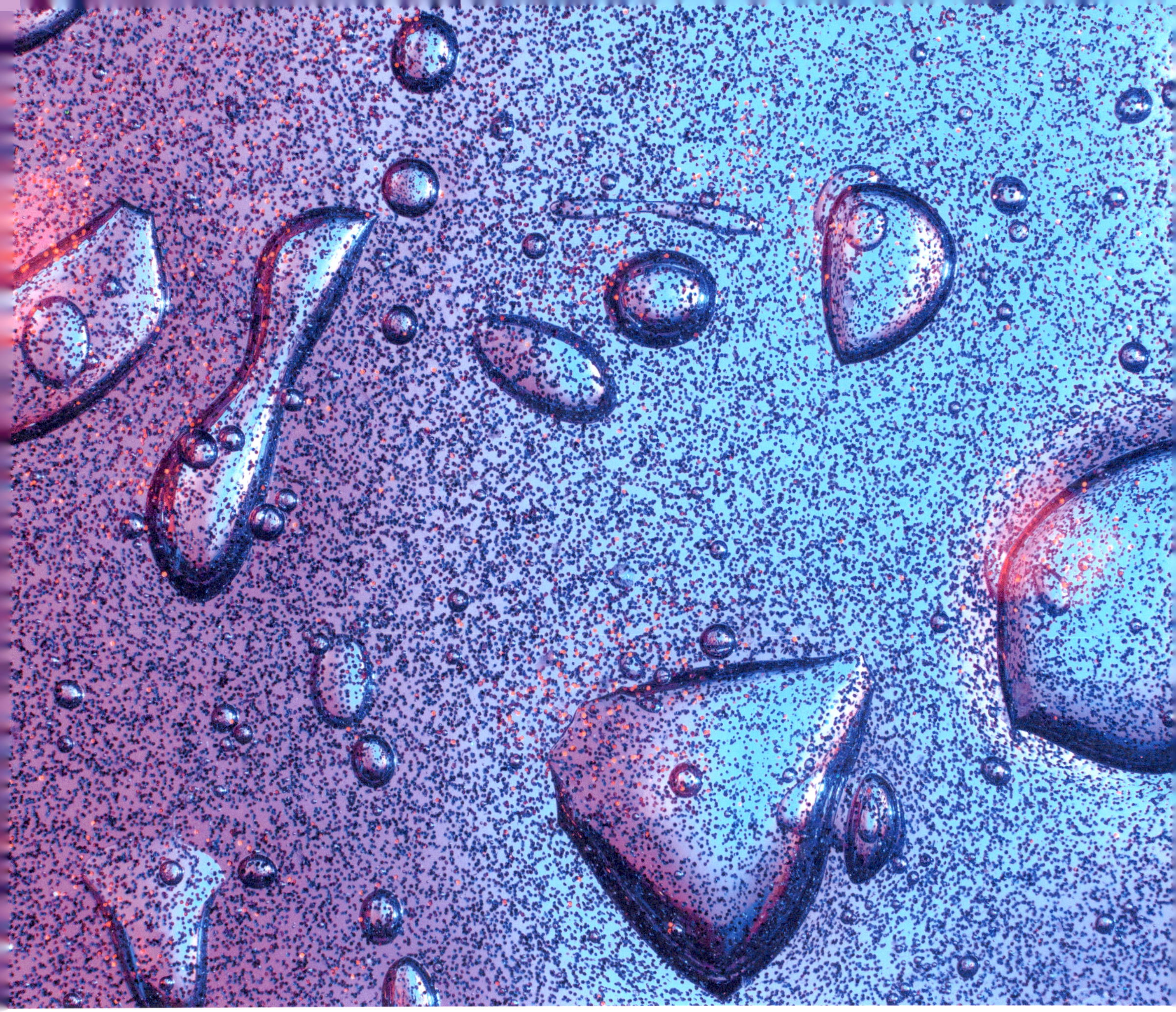

BUBBLE SLIME

DIRECTIONS

1. Grab a hunk of any slime and a plastic straw.
2. Cut your straw in half.
3. Stretch out a piece of slime into a flat shape and create a little sack to fasten around the end of the straw.
4. Make sure all the little gaps are filled, and hold this area tightly.
5. Blow into your straw as if you are blowing up a balloon and watch your bubble grow.

How big can you make it?

CRUNCHY SLIME

Special Note: Be sure to have large mixing bowls, measuring cups/spoons, something to mix with, and a tablecloth for your table. An apron or play clothes are appropriate. Gloves are optional.

INGREDIENTS

- **1 cup clear school glue**
- **1/2 cup water**
- **1/2 teaspoon borax dissolved in 1/2 cup warm water**
- **1 cup foam balls (any color or size)**

BACKGROUND

Crunchy slime is a popular variation of slime that incorporates foam balls or other small objects to create a unique texture and satisfying crunch when played with. The addition of foam balls adds a tactile element to the slime, giving it a crunchy sensation when squished and squeezed. This type of slime has gained popularity among slime enthusiasts due to its enjoyable sensory experience.

DIRECTIONS

1. In a bowl, mix the clear glue and water until well combined.
2. Slowly add the borax solution to the glue mixture while stirring continuously.
3. Keep stirring until the slime forms and starts to pull away from the sides.
4. Add the foam balls to the slime and mix until they are evenly distributed.
5. Knead the slime to incorporate the foam balls and achieve the desired texture.
6. Enjoy the satisfying crunchiness of your crunchy slime!

BOUNCY BALL

Special Note: Be sure to have large mixing bowls, measuring cups/spoons, something to mix with, and a tablecloth for your table. An apron or play clothes are appropriate. Gloves are optional.

INGREDIENTS

- **½ Cup Warm Water**
- **1 Tsp Borax Powder**
- **1 Tbs of Cornstarch**
- **1 Tbs Glue**
- **Food coloring**
- **2 small mixing cups**
- **Spoon or popsicle stick to mix with**

BACKGROUND

The concept of making bouncy balls revolves around polymers. Polymers are chains or links of molecules or compounds joined together. A chemical reaction takes place between the borax and the glue. The borax acts as a "cross-linker" to the polymer molecules in the glue - basically it creates chains of molecules that stay together when you pick them up. The cornstarch helps to bind the molecules together so that they hold their shape better.

DIRECTIONS

1. In a cup mix the warm water and the borax until it dissolves and label it "Borax Solution"
2. In another cup mix the glue and food coloring (3-4 drops) and label it "Ball Mixture."
3. Add 1/2 teaspoon of the borax solution and 1 tablespoon of cornstarch to the "Ball Mixture" cup. **DO NOT STIR!**
4. Allow the ingredients to interact on their own for 10-15 seconds and then stir them together to fully mix.
5. Once it becomes hard to stir, take it out of the cup and start molding the ball with your hands.
6. If the glue mixture is still sticky, squish it with your hands and dip it back in the water.Roll the mixture in your hands to make a ball. The more you handle the mixture the firmer it will become.

EXPLORE

Continue to explore by adjusting the amount of borax, glue, and cornstarch to get the highest bounce.

Also, experiment to discover the best way to get the bouncy ball to keep its bounce over time.

GALAXY SLIME

Special Note: Be sure to have large mixing bowls, measuring cups/spoons, something to mix with, and a tablecloth for your table. An apron or play clothes are appropriate. Gloves are optional.

INGREDIENTS

- **White or Clear PVA Washable School Glue**
- **Saline Solution**
- **Baking Soda**
- **Glow-in-the-dark paint/glue**
- **Glitter**
- **Food Coloring and Fun Mix-Ins**

DIRECTIONS

1. Mix 1/2 cup of glue and 1/2 cup of water in a bowl.
2. Add 1/2 tsp of baking soda and stir.
3. Add food coloring and mix-ins if desired.
4. Add 1 tbsp of saline solution.
5. Stir vigorously until mixture forms slimy ball.
6. Repeat with other colors

Make 4-5 small batches of the slime in different galaxy colors like black, orange, pink, blue and purple.

You can make a ½ batch of each to save on ingredients. Lay out your strips of slime in a row next to one another then twist into your galaxy slime.

EDIBLE SLIME

Special Note: Be sure to have large mixing bowls, measuring cups/spoons, something to mix with, and a tablecloth for your table. An apron or play clothes are appropriate. Gloves are optional.

INGREDIENTS

- **Cornstarch**
- **Package of Gelatin (Jello)**
- **Warm Water**
- **Cookie sheet or tray**

DIRECTIONS

1. Combine 1 cup of cornstarch and Jello powder together until fully mixed.
2. Add 1/4 cup of water and stir well
3. When the mixture becomes impossible to stir, add 1/4 cup more.
4. Stare kneading in 1 tbs of water at at time until the mixture can stretch or droop a bit. Add the water slowly so that you don't end up making Oobleck.

SALT DOUGH

INGREDIENTS

- **2 cups of all purpose flour**
- **1 cup of salt**
- **1 cup of water**
- **Cookie sheet or tray**
- **Cookie cutters (optional)**
- **Paint (optional)**

DIRECTIONS

1. In a large mixing bowl, stir together the flour and salt.
2. Gradually add water while stirring and mixing to form a dough with a Play-Doh like consistency.
3. Form the dough into a ball and knead for approximately 5 minutes, adding a bit more flour if the dough is too sticky or a bit more water if it's too dry.
4. Shape and decorate as desired

Note: If you want to bake your salt dough crafts, place them on a baking sheet in a 150-degree oven. Turn them frequently so that they dry out evenly. Baking time will vary depending on the size and thickness of your crafts, so just keep a close eye on them.

CHOCOLATE SLIME

Special Note: Be sure to have large mixing bowls, measuring cups/spoons, something to mix with, and a tablecloth for your table. An apron or play clothes are appropriate. Gloves are optional.

INGREDIENTS

- 1 cup chocolate spread or Nutella
- 1/2 cup powdered sugar
- 1 tablespoon cornstarch

DIRECTIONS

1. In a microwave-safe bowl, heat the chocolate spread for 30 seconds or until it is melted and smooth.
2. Add the powdered sugar and cornstarch to the melted chocolate spread.
3. Mix the ingredients until they are well combined and form a thick, dough-like consistency.
4. Let the mixture cool for a few minutes.
5. Knead the edible slime until it becomes smooth and stretchy.
6. Have fun playing with your delicious and safe-to-eat chocolate slime!

BACKGROUND

Chocolate slime is a creative twist on traditional slime recipes, where the slime is made to resemble the appearance and color of chocolate. It often uses ingredients like chocolate spread or Nutella to achieve a rich brown color and a pleasant chocolate scent. Chocolate slime is a fun and playful variation that appeals to both slime enthusiasts and chocolate lovers. It offers a unique sensory experience, combining the tactile nature of slime with the indulgent allure of chocolate. While chocolate slime is typically not meant for consumption, it provides a delightful sensory play option for slime enthusiasts.

BUBBLEGUM SLIME

Special Note: Be sure to have large mixing bowls, measuring cups/spoons, something to mix with, and a tablecloth for your table. An apron or play clothes are appropriate. Gloves are optional.

INGREDIENTS

- **1 cup white school glue**
- **1/2 cup water**
- **1/2 teaspoon baking soda**
- **1 tablespoon saline solution (containing borax)**
- **Bubblegum flavoring or extract**
- **Pink food coloring (optional)**

DIRECTIONS

1. In a mixing bowl, combine the white glue and water.
2. Stir in the baking soda until it is fully incorporated.
3. Add a few drops of bubblegum flavoring or extract and mix well.
4. If desired, add a few drops of pink food coloring and mix until the color is evenly distributed.
5. Gradually add the saline solution while stirring continuously.
6. Keep stirring until the slime comes together and becomes less sticky.
7. Knead the slime until it reaches the desired consistency, resembling bubblegum.
8. Enjoy the sweet scent and playful texture of your bubblegum slime!

BACKGROUND

Bubblegum slime offers a delightful sensory experience, combining the fun and familiar scent of bubblegum with the interactive and stretchy nature of slime. It appeals to those who enjoy both the tactile sensation of slime play and the sweet aroma of bubblegum.

BUBBLEGUM

SAND SLIME

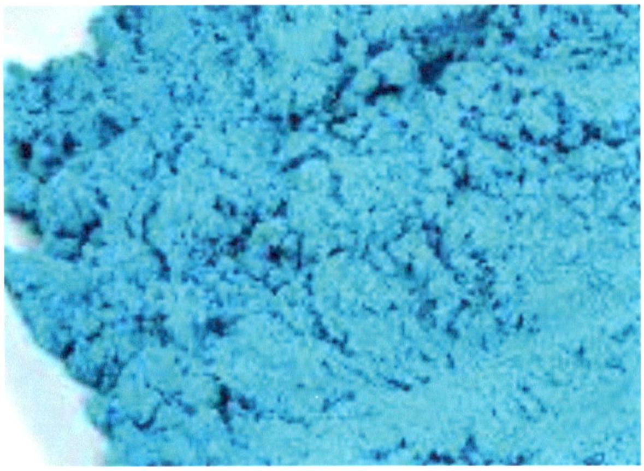

Special Note: Be sure to have large mixing bowls, measuring cups/spoons, something to mix with, and a tablecloth for your table. An apron or play clothes are appropriate. Gloves are optional.

INGREDIENTS

- **1 cup white school glue**
- **1 cup liquid starch**
- **Fine sand**

DIRECTIONS

1. Pour the white glue into a mixing bowl.
2. Gradually add the liquid starch while stirring continuously.
3. Keep stirring until the slime starts to form and becomes less sticky.
4. Sprinkle in fine sand, starting with small amounts, and mix well.
5. Continue adding sand until you achieve the desired texture and resemble sandy beach slime.
6. Knead the slime until it reaches the desired consistency.
7. Enjoy the tactile experience and the feeling of playing with sandy slime!

BACKGROUND

Sand slime combines the moldable nature of slime with the grainy texture of sand. It offers a unique sensory experience, allowing you to feel and shape the slime while also experiencing the sensation of playing with sand. Sand slime can evoke memories of the beach and provide a fun and tactile playtime

METALLIC SLIME

Special Note: Be sure to have large mixing bowls, measuring cups/spoons, something to mix with, and a tablecloth for your table. An apron or play clothes are appropriate. Gloves are optional.

INGREDIENTS

- **1/2 cup clear school glue**
- **1/2 cup liquid starch**
- **Metallic acrylic paint or metallic food coloring**

DIRECTIONS

1. Pour the clear glue into a mixing bowl.
2. Add a few drops of metallic acrylic paint or metallic food coloring and mix well until the color is evenly distributed.
3. Gradually add the liquid starch while stirring continuously.
4. Keep stirring until the slime starts to form and becomes less sticky.
5. Knead the slime until it reaches the desired consistency, and the metallic shine is visible.
6. Enjoy the dazzling and shiny effect of your metallic slime!

BACKGROUND

Metallic slime adds a touch of glamour and shine to the slime experience. The metallic pigments create a lustrous and eye-catching appearance, making the slime visually captivating and exciting to play with

CLOUD SLIME

Special Note: Be sure to have large mixing bowls, measuring cups/spoons, something to mix with, and a tablecloth for your table. An apron or play clothes are appropriate. Gloves are optional.

INGREDIENTS

- **1 cup white school glue**
- **1/2 cup instant snow powder**
- **1/2 cup water**
- **1/2 teaspoon baking soda**
- **1 tablespoon saline solution (containing borax)**
- **Cotton candy scent (optional)**

DIRECTIONS

1. In a mixing bowl, combine the white glue and instant snow powder.
2. Mix well until the snow powder is fully incorporated and the mixture resembles a fluffy cloud texture.
3. In a separate container, mix the water and baking soda until the baking soda dissolves.
4. Add a few drops of cotton candy scent to the water mixture and mix well (optional).
5. Gradually pour the water mixture into the glue and snow mixture while stirring continuously.
6. Add the saline solution and continue stirring until the slime comes together and becomes less sticky.
7. Knead the slime until it reaches the desired consistency and resembles a soft and fluffy cloud.
8. Enjoy the pillowy and airy texture of your cloud slime!

BACKGROUND

Cloud slime is a type of slime that has a light and airy texture, reminiscent of fluffy clouds. The addition of instant snow powder gives it a unique and satisfying squishy feel. Playing with cloud slime can evoke a sense of wonder and imagination as it resembles the softness and fluffiness of clouds.

JELLY SLIME

Special Note: Be sure to have large mixing bowls, measuring cups/spoons, something to mix with, and a tablecloth for your table. An apron or play clothes are appropriate. Gloves are optional.

INGREDIENTS

- **White or Clear PVA Washable School Glue**
- **Fragrance oil or Gelatin powder in any flavor (Sugar-free gelatin not recommended)**
- **Food Coloring and Fun Mix-Ins**

DIRECTIONS

1. Choose your basic slime recipe: Borax, Liquid starch, or Saline solution
2. If using Gelatin powder: use a ½ tsp of baking soda and 1 ½ tablespoons of saline solution.

- Add 1 teaspoon of any flavored powdered gelatin after the glue and water have been mixed together. Food coloring can be added but is not necessary. Stir

3. If using fragrance oil: During the mix-ins step, add a drop or two of fragrance oil and continue with the basic slime recipe directions.
4. Add food coloring and mix-ins if desired.
5. Add your activator
6. Stir vigorously until mixture forms slimy ball.

BACKGROUND

Jelly slime is a playful variation that mimics the wobbly and jiggly texture of jelly. By incorporating jelly powder mix, you can give the slime a burst of color and flavor, making it a sensory delight.

RAINBOW SLIME

Special Note: Be sure to have large mixing bowls, measuring cups/spoons, something to mix with, and a tablecloth for your table. An apron or play clothes are appropriate. Gloves are optional.

INGREDIENTS

- **White or Clear PVA Washable School Glue**
- **Other ingredients based on basic slime recipe**
- **Glitter**
- **Food Coloring and Fun Mix-Ins**

DIRECTIONS

1. Choose your basic slime recipe: Borax, Liquid starch, or Saline solution
2. Add food coloring and mix-ins if desired.
3. Add your activator
4. Stir vigorously until mixture forms slimy ball
5. Repeat with the 5 other colors of the rainbow
6. To make the rainbow, lay the colors out in a strip. Pick up the top of the strip and twist!

Note: The colors will eventually mix together leaving a muddy color.
You can make a ½ batch of each to save on ingredients. Lay out your strips of slime in a row next to one another then twist into your galaxy slime.

PRIMARY MIX SLIME

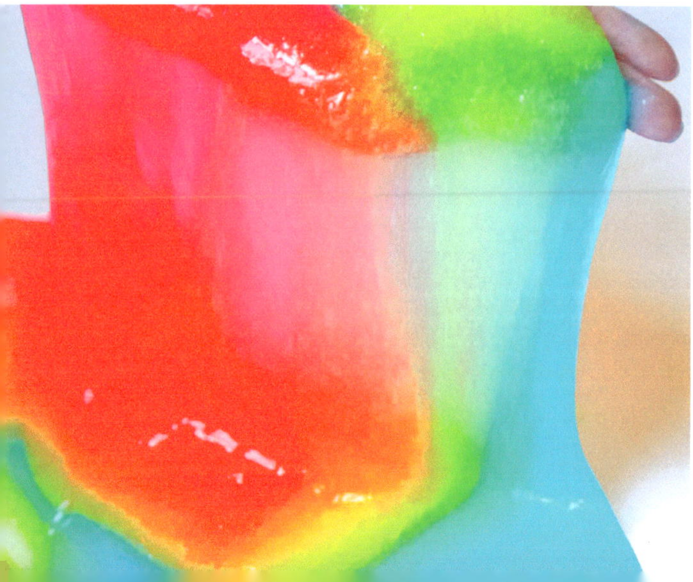

DIRECTIONS

1. Create the three primary colors: red, yellow, and blue.
2. Layer them in a glass cup or pan.

Watch the secondary colors emerge where the slime colors meet. What happens if you leave it overnight?

While making slime is a fun and creative activity, it can sometimes lead to some common problems. In this chapter, we'll cover some of the most common problems that may arise when making slime and provide solutions for each one.

STICKY SLIME

Sticky slime is a common problem that can occur when making slime. Sticky slime can be frustrating to play with and difficult to manipulate. Here are some solutions for fixing sticky slime:

Add more activator: If your slime is too sticky, try adding more activator, such as borax or saline solution. Start with a small amount and work it into the slime until it becomes less sticky.

Knead the slime: Kneading the slime can help to work the activator into the slime and make it less sticky. Keep kneading until the slime becomes smooth and stretchy.

The science behind sticky slime: Sticky slime is usually the result of not adding enough activator to the glue. The activator is what causes the polymer chains in the glue to link together and form the slime. Without enough activator, the slime may remain sticky and difficult to play with.

CLEANING UP A SLIMY MESS

Slime Removal Tip #1

Remove as much of the slime as possible from the material. Squirt dish soap onto the slimy area. Scrub the area under running water. Once the majority of the slime is gone, you can wash as usual on laundry day!

Slime Removal Tip #2

Remove as much of the slime as possible from the material. Pour white vinegar onto the affected area and use a clean scrub brush to remove residue. Add more vinegar as needed and continue scrubbing. Rinse. Wash in the laundry as usual.

SLIME

camp

It's time for some slimy fun this summer! Fluffy slime, magnetic slime, glow-in-the-dark slime, color-changing slime, you name it... we're making it!

Visit our website www.achievesuccesstutoring.com to find out more information on our camps!

Made in the USA
Las Vegas, NV
07 November 2026

57650032R00021